Abdelhafid Mimouni

A dive into the labyrinth of Schottky defects

Abdelhafid Mimouni

A dive into the labyrinth of Schottky defects

Imprint

Any brand names and product names mentioned in this book are subject to trademark, brand or patent protection and are trademarks or registered trademarks of their respective holders. The use of brand names, product names, common names, trade names, product descriptions etc. even without a particular marking in this work is in no way to be construed to mean that such names may be regarded as unrestricted in respect of trademark and brand protection legislation and could thus be used by anyone.

Cover image: www.ingimage.com

This book is a translation from the original published under ISBN 978-620-6-71224-4.

Publisher:
Sciencia Scripts
is a trademark of
Dodo Books Indian Ocean Ltd. and OmniScriptum S.R.L publishing group

120 High Road, East Finchley, London, N2 9ED, United Kingdom
Str. Armeneasca 28/1, office 1, Chisinau MD-2012, Republic of Moldova, Europe
Printed at: see last page
ISBN: 978-620-7-63567-2

"A dive into the labyrinth of Schottky defects".

Author: Dr. Abdelhafid Mimouni is an independent researcher specializing in bioinorganic systems chemistry, with extensive expertise in macromolecular synthesis and characterization. He obtained his PhD in Chemistry from the University of Paris XII "1997" and a Diplôme des Etudes Approfondies des systèmes bioinorganiques from the University of Paris XI "1993".

Summary:

Schottky defects, discovered by Walter H. Schottky in 1938, are crystalline imperfections that play a crucial role in the properties of materials. These defects, which result from the loss of atoms in a crystalline structure, affect the electrical conductivity, mechanical strength, optical and catalytic properties of materials. Understanding and controlling these defects are major challenges in many fields of science and technology, including materials engineering, electronics, catalysis and emerging technologies such as quantum computing and photonic devices. Current research is exploring the potential applications of Schottky defects, while also seeking to develop new materials and techniques to optimize their use in various fields of application.

Plan :

Introduction

In the vast field of materials science, the importance of crystalline defects is indisputable. Among these imperfections, Schottky defects occupy a special place, offering serious opportunities for scientific and technological exploration. This book is a deep dive into the intriguing world of Schottky defects, from their origins to their potential applications, offering a comprehensive overview of their role in the design and optimization of modern materials.

To fully understand the impact of Schottky defects, it's essential to go back to the roots of their discovery. Wilhelm Schottky, a German physicist and engineer born in 1886, was one of the pioneers in the study of crystalline defects. Throughout his remarkable career, Schottky made significant contributions to various fields of physics, but it was his work on defects in crystals that won him lasting recognition in the scientific community.

Born in Zurich, Switzerland, Schottky studied physics at the University of Munich, where he obtained his doctorate in 1912. He subsequently held various academic positions, including Professor of Physics at the University of Rostock and the University of Wrocław. Schottky made important contributions to theoretical and experimental physics, focusing on material properties and crystalline defects.

Among his many achievements, Schottky is best known for his work on defects in crystals, in particular the point defects that today bear his name. Schottky defects, characterized by the presence of empty spaces in the crystal structure, have fascinated scientists and engineers ever since for their role in the properties of materials.

Today, Schottky defects continue to occupy a central place in materials science research. Understanding and manipulating them offers unprecedented opportunities to design tailor-made materials with specific properties, paving the way for a new era of discovery and innovation in fields ranging from electronics to energy and beyond.

This book is a journey through the exciting landscape of Schottky defects, offering an in-depth exploration of their history, nature and impact on our modern world. Whether you're a seasoned researcher, curious student or industry professional, we hope this dive into the world of Schottky defects will awaken your fascination with the infinite possibilities they offer in the quest for advanced materials and technologies.

Chapter 1: General introduction

Introduction

Schottky defects, point-like lacunae in crystal lattices, are of vital importance in various fields of technology. They form when atoms and their associated ions are absent from adjacent sites, creating a pair of linked gaps. These defects significantly influence the physical, chemical and electronic properties of many materials, such as electrical conductivity, mechanical strength, catalytic performance and optical properties. A detailed understanding of their formation, evolution and impact is essential for optimizing materials performance and developing innovative new applications.

1.1   Definition of Schottky defects

Schottky defects, at the atomic level, result from the simultaneous absence of an atom and its associated ion at adjacent sites in the crystal lattice. This absence induces a local disturbance in the lattice, modifying the material's properties. Their formation is governed by thermodynamic principles designed to minimize the free energy of the system.

1.2   Material ageing and Schottky defect evolution

Over time, the concentration and distribution of Schottky defects change in response to operating conditions (temperature, stress, environment, etc.). This intrinsic aging of materials can alter their physical and chemical properties. For example, at high temperatures, Schottky defects can recombine or diffuse, resulting in microstructural changes that affect mechanical strength, electrical conductivity and other important properties.

1.3   Objectives of the book

The aim of this book is to provide a comprehensive overview of Schottky defects, from their origin to their influence on material properties, as well as strategies for controlling them. The book will cover fundamental principles, characterization techniques, theoretical models and current challenges in this field.

1.4 Examples of the impact of Schottky defects in real applications

1   High-temperature alloys: In jet engines, materials have to withstand extreme temperatures. Schottky defects can play a crucial role in element diffusion and microstructure stabilization at high temperatures, helping to maintain the mechanical properties and durability of engine components.

2   Anti-corrosion coatings: Aircraft are exposed to corrosive environments, such as humidity and chemicals on runways or during high-altitude flights. Schottky defects can be used to design protective coatings that exploit specific ionic diffusion properties to prevent corrosion of metal surfaces.

3.   composite materials : Composite materials are increasingly used in aircraft construction to reduce weight and improve performance.

Schottky defects can influence the interfacial bonding between fibers and matrix, affecting the mechanical strength and durability of composites.

4   Catalysts for propulsion: Schottky defects can be exploited in the design of catalysts for aeronautical propulsion systems, such as jet engines. By modifying surface properties and chemical reactivity, these catalysts can improve energy efficiency and reduce pollutant emissions.

5.   materials for aerodynamic structures: Schottky defects can affect the mechanical properties of materials used in the construction of aerodynamic structures, such as wings and fuselages. By understanding and controlling these defects, it is possible to optimize the strength, lightness and durability of these structures, thus contributing to aircraft safety and efficiency.

1.5 Links between Schottky defects and dislocations in material deterioration

Schottky defects and dislocations both play an essential role in the degradation and ageing mechanisms of materials. Although different in nature, these two types of crystalline defects interact and influence each other, contributing to the evolution of material properties during use.

On the one hand, the presence of Schottky defects locally modifies the lattice structure, creating zones of distortion and disorder. These disturbances encourage the nucleation and migration of dislocations, which are linear defects corresponding to relative displacements of atomic planes.

Dislocations move more easily in regions of high Schottky defect density, as these point gaps act as trapping or nucleation sites for dislocations. In turn, the movement and accumulation of dislocations lead to changes in the local concentration of Schottky defects, altering their scattering and energy properties.

This complex interplay between point and linear defects plays a major role in the fatigue, creep and corrosion processes of materials. For example, the increased mobility of dislocations, facilitated by Schottky defects, promotes the formation of microcracks and crack propagation, accelerating the mechanical degradation of the material.

Similarly, the combined presence of these two types of defect can modify the transport properties of chemical species (ions, electrons, etc.), influencing oxidation, passivation or corrosion reactions on the surface or in the volume of the material. The result is a progressive deterioration in the material's performance and service life over the course of its use.

Chapter 2. Fundamentals of Schottky defects

Schottky defects are point-like gaps in the crystal lattices of materials, which play an essential role in various fields, including aerospace. An in-depth understanding of the fundamentals of these defects is crucial for assessing the stability and durability of materials used in aircraft construction. This chapter explores the atomic structure of Schottky defects, the mechanisms of their formation and the associated thermodynamic implications.

2.1 Atomic structure of Schottky defects and interactions with dislocations

At the atomic level, a Schottky defect is characterized by the simultaneous absence of an atom and its associated ion at adjacent sites in the crystal lattice. This pair of linked gaps induces a local distortion of the lattice, affecting the material's physical and mechanical properties. This distortion can be measured using various techniques such as X-ray diffraction and neutron scattering spectroscopy. For example, devices such as laser interferometers monitor in real time variations in the length or deformation of a material subjected to mechanical loads.

Schottky defects and dislocations are two types of crystal defect that interact closely in the deterioration of materials. The presence of Schottky defects creates zones of distortion and disorder in the crystal lattice, favoring the nucleation and migration of dislocations. Dislocations, in turn, move more easily into regions rich in Schottky defects, modifying the local concentration of these defects and influencing their diffusion and energy properties. This complex interplay between point and linear defects plays a major role in the fatigue, creep and corrosion processes of materials.

In the field of aviation, where material strength and durability are crucial, understanding these interactions between Schottky defects and dislocations is essential for assessing material performance under a wide range of flight conditions.

One of the equations commonly used to quantify the mechanical distortion of a material is Hooke's law, expressed as: $\sigma = E\sigma\, \epsilon$

Where:

-σ : is the stress applied to the material (in Pa),

-E: is the Young's modulus of the material (in Pa),

-ϵ: is the deformation or stress defined as the relative change in length of the material with respect to its initial length.

This equation enables us to understand how the stress applied to a material is related to its deformation, which is crucial for assessing the mechanical response of materials to external loads.

2.2 Schottky defect formation mechanisms

The formation of Schottky defects is a process regulated by complex thermodynamic and kinetic mechanisms. At elevated temperatures, the probability of the formation of point defects is increased by the entropy of the system. This increase in entropy promotes the disorganization of the crystal lattice, enabling the creation of linked pairs of vacancies characteristic of Schottky defects.

For example, during rapid cooling after heat treatment, concentration gradients of atoms can form in the material. These concentration gradients create conditions conducive to the formation of vacancies, as atoms can move faster than their recombination mobility. As a result, Schottky defects can form at specific locations where vacancies are more likely to meet and stabilize.

Once formed, these Schottky defects can play a critical role in material degradation. They act as stress concentration points, promoting the initiation and propagation of cracks in the material. Indeed, the presence of these defects locally weakens the crystalline structure, making the material more vulnerable to external stresses. A detailed understanding of the mechanisms by which Schottky defects form is therefore essential for designing manufacturing and heat treatment processes that minimize the formation of these undesirable defects.

By focusing on understanding the mechanisms underlying the formation of Schottky defects, engineers and researchers can develop strategies for designing stronger, more durable materials. This may include optimizing heat treatment parameters, developing new Schottky defect-resistant alloys, or implementing surface coating techniques to enhance resistance to the stresses induced by these defects.

In short, in-depth analysis of Schottky defect formation mechanisms is crucial to the design of high-performance materials in a variety of technological applications. By understanding the conditions and processes that favor the formation of these defects, it becomes possible to control them and minimize their impact on material performance.

2.3 Thermodynamics of Schottky defects

The thermodynamics of Schottky defects is a crucial discipline for understanding their behavior at different temperatures, pressures and compositions. It is based on thermodynamic equations and fundamental principles that describe the reactions and equilibria involving these point defects in crystalline materials.

The equilibrium concentration of Schottky defects is closely dependent on parameters such as temperature, pressure and material composition. This dependence is often described by thermodynamic equations such as the Gibbs-Duhem equations, which describe variations in the free enthalpy of a system in equilibrium.

At high temperatures, the entropy of the system favors the formation of Schottky defects. This is because, as the temperature rises, the atoms in the crystal lattice acquire greater kinetic energy, increasing their mobility and facilitating the formation of point-like vacancies. Thus, at higher temperatures, the concentration of Schottky defects in the material is generally higher.

However, at lower temperatures, reverse reactions, such as defect migration and recombination, may predominate. These migration and recombination processes can lead to a decrease in the concentration of Schottky defects in the material.

In aviation, the thermodynamics of Schottky defects is of paramount importance in predicting the behavior of materials in extreme environments. By understanding the thermodynamic principles governing the formation and reaction of these defects, engineers can design stronger, more durable materials for aeronautical applications. For example, by adjusting heat treatment parameters or selecting specific alloys, it is possible to control the concentration of Schottky defects in the material, which can have a significant impact on its mechanical strength and durability.

Integrating these principles into material design helps to ensure the safety and reliability of aeronautical structures, while optimizing their mechanical performance and corrosion resistance. By combining an in-depth understanding of Schottky defect thermodynamics with advanced modeling techniques, engineers can anticipate and prevent potential material failures, contributing to the safety and efficiency of aeronautical systems.

Chapter 3: Impact of Schottky defects on material properties

Schottky defects, as point-like gaps in crystal lattices, have a significant influence on a variety of material properties, including electrical, mechanical, optical and catalytic.

3.1 Effects on electrical properties

Schottky defects exert a significant impact on the electrical conductivity of materials by altering the mobility of charge carriers within the crystal lattice. Their effects on electrical properties depend on a variety of factors, including the type of material, the density of the defects, and the nature of the interactions between the charge carriers and the Schottky defects themselves.

When Schottky defects are present in a material, they can affect its electrical conductivity in different ways:

1   Increased conductivity: In some cases, the presence of Schottky defects can enhance electrical conductivity by creating alternative conduction paths for charge carriers. This can occur when Schottky defects act as scattering centers, facilitating the movement of charge carriers through the material.

2)   Reduced conductivity: Conversely, in other situations, Schottky defects can lead to a reduction in electrical conductivity by disrupting the mobility of charge carriers. For example, the presence of defects can interfere with the movement of electrons or holes in the material, reducing its overall conductivity.

Schottky defects can also act as doping sites in semiconductor materials. When combined with impurities intentionally introduced into the material, these defects can influence doping levels and thus the electrical properties of the material. For example, the presence of Schottky defects can alter the concentration of free charge carriers, thus affecting the material's electrical conductivity.

The relationship between Schottky defects and the electrical properties of materials can be described by various equations and models, including Ohm's law for conducting materials and transport equations for semiconductors. These models take into account defect density, charge carrier mobility and other parameters to predict the overall electrical behavior of the material.

In conclusion, understanding the effects of Schottky defects on the electrical properties of materials is essential for the design and optimization of electronic devices and semiconductor components. By integrating this knowledge into materials and device design, it is possible to fully exploit the advantages or minimize the detrimental effects of these defects on the electrical performance of systems.

3.2 Effects on mechanical properties

Schottky defects have a significant influence on the hardness, ductility and resistance to mechanical stress, and therefore on the durability of materials. Their presence disturbs the crystalline structure, which can have various effects on the mechanical properties of materials.

1   Hardness: Schottky defects can affect the hardness of materials by creating obstacles to plastic deformation. When stress is applied to the material, these defects can act as anchor points for dislocations, hindering their movement and increasing resistance to deformation. This usually results in an increase in the material's hardness.

2   Ductility: By disrupting the crystal structure, Schottky defects can also affect the ductility of materials. When Schottky defects are present, dislocation propagation can be impeded, limiting the material's ability to deform plastically before fracture. Consequently, the presence of Schottky defects can reduce the ductility of the material, making it more brittle.

3   Resistance to mechanical stress: Schottky defects act as stress concentration points in the material. When external stress is applied, these defects can act as starting points for crack initiation and propagation, thereby reducing the material's overall resistance to mechanical stress. Thus, the presence of Schottky defects can compromise the strength and durability of materials, particularly under severe loading conditions.

The relationship between Schottky defects and the mechanical properties of materials can be described by models such as plasticity theory or fracture mechanics. These models take into account parameters such as defect density, grain size and the nature of interactions between defects and dislocations to predict the overall mechanical behavior of the material.

In conclusion, understanding the effects of Schottky defects on the mechanical properties of materials is essential for designing robust and durable structures and components. By identifying and quantifying the impact of these defects on the strength and durability of materials, it is possible to develop design strategies aimed at minimizing the adverse effects of these defects and improving the mechanical performance of materials in a range of industrial and technological applications.

3.3 Effects on optical properties

Schottky defects exert a significant influence on the optical properties of materials, affecting their ability to absorb, emit and scatter light. This influence is particularly crucial in applications such as optoelectronics, photonics and optical devices.

1   Light absorption and emission: Schottky defects can modify the absorption and emission bands of materials. By disrupting the regularity of the crystal lattice, these defects create additional energy levels in the material's electronic structure. These energy levels can affect electronic transitions, leading to changes in the material's absorption and emission spectra. For example, Schottky defects can introduce new energy levels

into the band gap of a semiconductor material, altering its optical and electrical properties.

2 Transmittance and reflectance: The presence of Schottky defects can also influence the transmittance and reflectance of materials. These defects can act as light-scattering sites, resulting in increased scattering and decreased transmittance through the material. In addition, Schottky defects can alter the refractive index of the material, affecting its ability to reflect incident light. Consequently, the presence of Schottky defects can alter the overall optical properties of the material, with important implications for the design of optical and photonic devices.

The relationship between Schottky defects and the optical properties of materials can be studied using techniques such as optical spectroscopy, absorption spectroscopy, photoluminescence and light scattering. These techniques can be used to characterize the absorption and emission spectra of materials, as well as their overall optical behavior in the presence of Schottky defects.

In conclusion, the presence of Schottky defects can have a significant impact on the optical properties of materials, necessitating a thorough understanding of their influence in various optical and photonic applications. By integrating this knowledge into the design of new optical materials and devices, the effects of Schottky defects can be fully exploited to develop innovative, high-performance optical technologies.

3.4 Effects on catalytic properties

Schottky defects exert a significant influence on the catalytic properties of materials, playing a crucial role in many heterogeneous catalysis processes. Their impact on catalytic efficiency depends on a number of factors, such as the nature of the chemical reactions, the composition of the materials and the structure of the defects themselves.

1   Modification of the active surface: Schottky defects can modify the active surface available for chemical reactions by disturbing the regularity of the crystal lattice at the surface of the material. These defects create binding sites with the reactants, increasing the probability of formation of intermediate reaction species and facilitating reaction steps. For example, in metal-based catalysts, Schottky defects can act as active sites for the adsorption and dissociation of reactive molecules, enhancing the overall catalytic efficiency of the material.

2.   influence on reaction kinetics : The presence of Schottky defects can also affect the kinetics of catalytic reactions by modifying the energy barriers associated with the various stages of the reaction process. For example, Schottky defects can facilitate the desorption of reaction products, thereby accelerating the overall reaction rate. Moreover, these defects can also modify reaction selectivity by favoring certain reaction pathways over others, which can have a significant impact on the yield and selectivity of catalytic products.

3.   examples of impact on applications : Schottky defects play a crucial role in many catalytic applications, such as the conversion of synthesis gas to liquid fuels, the reduction of nitrogen oxides in automotive catalytic converters, and the production of hydrogen from methane reforming. In these systems, the presence of Schottky defects in metal or ceramic catalysts can significantly improve catalytic activity, product selectivity and the long-term stability of the catalytic material.

In conclusion, Schottky defects play a crucial role in the catalytic properties of materials, influencing the efficiency, selectivity and stability of heterogeneous catalysts in a variety of industrial and environmental applications. A thorough understanding of their impact on catalytic processes is essential to design and optimize new catalytic materials with improved performance and enhanced durability.

Chapter 4: Identification and characterization of Schottky defects

Accurate understanding and characterization of Schottky defects is essential to assess their impact on material properties and develop prevention and optimization strategies.

4.1 Experimental techniques

Experimental techniques play an essential role in identifying and characterizing Schottky defects at different scales, from microscopic to atomic. Among these techniques, electron microscopy plays a key role. Scanning electron microscopy (SEM) and transmission electron microscopy (TEM) allow the crystalline structure of materials to be observed with sub-nanometric resolution, enabling Schottky defects to be visualized on an atomic scale.

In addition to electron microscopy, other experimental techniques such as neutron scattering spectroscopy, X-ray diffraction and Raman spectroscopy are also used to study Schottky defects. These techniques make it possible to determine the spatial distribution, density and structural properties of defects in materials, providing comprehensive information on their nature and behavior.

In aviation, these experimental techniques are of vital importance in assessing the quality and reliability of materials used in the construction of aeronautical structures. By combining these different methods, engineers can comprehensively characterize the Schottky defects present in materials, enabling the design of safer and more durable aeronautical components. Electron microscopy, in particular scanning electron microscopy (SEM) and transmission electron microscopy (TEM), offers high spatial resolution for direct visualization of Schottky defects at the atomic scale. These techniques enable us not only to observe the distribution and density of defects, but also to characterize their structure and morphology.

Spectroscopy, such as X-ray absorption spectroscopy, Raman spectroscopy and photoelectron spectroscopy (XPS), is widely used to probe the electronic and chemical properties of Schottky defects. By analyzing electronic transitions and photon-matter interactions, these techniques provide valuable information on the nature and state of defects present in materials.

4.2 Modelling and numerical simulations

Modeling and numerical simulations are essential tools for understanding the mechanisms by which Schottky defects form and evolve in materials. These approaches enable us to simulate the behavior of materials at the atomic scale and predict their response under different environmental and processing conditions.

Atomistic simulations, such as molecular dynamics and the first-principles method, enable the movement of atoms and defects in materials to be modeled with great precision. They provide detailed information on the structure, energy and stability of Schottky defects, enabling us to understand their effects on material properties.

Atomistic modeling, based on ab initio computational methods such as DFT (Density Functional Theory), enables the properties of Schottky defects to be predicted at the atomic scale. By simulating the interactions between atoms and electrons, these approaches provide valuable insights into the structure, energy and stability of defects, as well as their effects on material properties.

Numerical simulations on the mesoscopic scale, such as molecular dynamics and finite element modeling, enable us to study the behavior of Schottky defects on larger scales. They take into account the interactions between defects, dislocations and external stresses, enabling us to predict the macroscopic behavior of materials containing Schottky defects.

At the same time, mesoscopic and macroscopic simulations, such as phase field theory methods and finite element simulations, allow us to model the overall behavior of materials, taking into account the presence of Schottky defects. They analyze the impact of defects on the mechanical, electrical, optical and catalytic properties of materials, providing valuable information for the design and optimization of materials in various applications.

In aviation, numerical modeling and simulations play a crucial role in the development of stronger, lighter and more durable materials for aeronautical structures. Integrating these approaches into the design process enables engineers to optimize material performance by minimizing the impact of Schottky defects, thus ensuring the safety and reliability of aeronautical components.

4.3 Links between experimental and theoretical observations

Establishing links between experimental observations and theoretical predictions is essential for validating models and interpreting results in a meaningful way. Multi-scale correlation techniques, such as TEM-STEM (Transmission Electron Microscopy-Scanning Transmission Electron Microscopy) correlation, make it possible to link microscopic observations at the atomic scale with the macroscopic behavior of materials. This integration of experimental data and modeling results provides a better understanding of the formation, stability and effects of Schottky defects on material properties, paving the way for the development of new materials and technologies.

The correlation between experimental observations and theoretical models is crucial to validate results and better understand the underlying processes. These links help improve the accuracy of predictions and guide the development of new materials with optimized properties.

In conclusion, the identification and characterization of Schottky defects are essential to understanding their impact on material properties and developing strategies to optimize their performance in various applications.

Chapter 5. Strategies for controlling and reducing Schottky defects

Schottky defects, because of their impact on material properties, require a systematic approach to their control and reduction. This chapter explores various methods and techniques aimed at achieving this goal, as well as their practical applications.

5.1 Material synthesis and processing methods

The synthesis of materials free of Schottky defects and processing to reduce their occurrence are major objectives in many areas of research and application. This section examines in detail the various methods of material synthesis, such as controlled crystal growth and chemical synthesis, as well as processing techniques, including heat treatments and purification methods. Understanding the mechanisms of Schottky defect formation enables the design of effective strategies to minimize their occurrence and optimize material properties.

In addition to traditional methods, innovative approaches such as laser-assisted synthesis and nanofabrication enable the structure and purity of materials to be controlled with great precision, thereby reducing the formation of Schottky defects. For example, in the field of magnetic materials, the manufacture of multilayer thin films by vacuum deposition enables precise control of the magnetic orientation and avoids the formation of Schottky defects that could disrupt the magnetic properties of the materials.

In addition, the use of vacuum deposition and directed crystallization techniques offers ways of producing materials with reduced defect density, thus improving their stability and performance. For example, in the field of semiconductors, epitaxial growth enables the production of high-quality crystals with a well-ordered atomic structure, minimizing the formation of Schottky defects that could affect the material's electronic properties.

In aviation, where material reliability and durability are crucial, these synthesis and processing methods are of particular importance. By precisely controlling the structure and purity of materials, engineers can minimize Schottky defects and guarantee optimum performance under rigorous environmental conditions.

5.2 Schottky defect engineering: Modulation of material properties

Controlled doping of materials offers a promising route to selectively modify electronic and structural properties, while influencing the density and distribution of Schottky defects. This section explores different doping techniques, such as ion and chemical doping, as well as defect engineering strategies to control the formation and effect of point defects. By fine-tuning doping parameters and growth conditions, it is possible to modulate material properties and reduce the impact of Schottky defects on performance.

In semiconductors, doping is widely used to control electrical conductivity and modify energy bands, which can influence the formation and reactivity of Schottky defects. For example, n- or p-type doping in semiconductors can modify the charge carrier density and thus the mobility of electrons or holes, which can influence the formation and stability of Schottky defects.

Approaches such as selective doping and sequential doping make it possible to precisely control the concentration and location of Schottky defects, offering an effective means of regulating material properties. For example, in silicon-based solar cells, selective doping can be used to create p and n regions in a controlled manner, minimizing the formation of Schottky defects at contact interfaces, thereby improving solar energy conversion efficiency.

In addition, nanoscale defect engineering can be used to create customized hardware architectures, minimizing the undesirable effects of Schottky defects on overall performance. For example, in graphene-based electronic devices, defect engineering can be used to control electron transport properties, thereby improving device efficiency and reliability.

In aviation and other fields where material reliability and performance are crucial, controlled doping and defect engineering offer promising ways of optimizing material properties and reducing the adverse effects of Schottky defects. By understanding and exploiting these strategies, engineers can design better performing and more durable materials for a range of critical applications.

5.3 Practical applications of Schottky fault monitoring

Understanding and mastering Schottky defects opens the way to many practical applications in various fields. This section highlights how knowledge of Schottky defects can be applied in real-life contexts, with an illustrative example on the prevention of galvanic corrosion in metal structures. By using Schottky defect control strategies, it is possible to extend the life of materials, improve their performance and foster the development of innovative, sustainable technologies.

For example, in the field of solar energy, knowledge and control of Schottky defects are essential for improving the efficiency of solar cells. By minimizing the formation of defects at the interface between layers of semiconductor materials, it is possible to optimize the capture and conversion of solar energy, helping to increase the overall efficiency of solar panels.

Similarly, in the field of catalysis, the manipulation of Schottky defects can be used to optimize the performance of catalysts in chemical reactions. By controlling the density and distribution of defects on the surface of catalysts, it is possible to improve their catalytic activity, selectivity and stability, paving the way for new applications in clean energy production and green chemistry.

Finally, in the aerospace field, Schottky defect control is crucial for strengthening the materials used in aircraft and satellite construction. By minimizing the formation of defects within materials, it is possible to improve their mechanical strength, durability and corrosion resistance, helping to guarantee the safety and reliability of aerospace structures in extreme environments.

By using Schottky defect control approaches, it is possible to meet the growing need for functional and sustainable materials in a wide range of industries and application areas. By fully exploiting the potential of Schottky defects, researchers and engineers can help accelerate innovation and develop more efficient and sustainable technologies to meet the challenges of our time.

Chapter 6. Challenges and prospects

This chapter explores current challenges in understanding Schottky defects, as well as promising prospects for future research and potential new applications for these defects.

6.1 Current limitations in the understanding of Schottky defects

Despite the progress made in characterizing and modeling Schottky defects, there are still significant limitations in our understanding of these point defects. The mechanisms by which Schottky defects form and interact with other crystalline defects, such as dislocations, often remain poorly understood. Moreover, the variability of Schottky defect properties as a function of material growth and processing conditions is a major challenge to overcome.

Future research should therefore focus on a deeper understanding of these aspects, exploring the complex interactions between Schottky defects and other crystal imperfections, as well as developing more accurate models to predict their behavior under different conditions.

6.2 Emerging research areas

Several emerging areas of research offer new perspectives for deepening our understanding of Schottky defects, and pave the way for significant advances in materials design and engineering.

Nanotechnology, for example, represents a promising field for the controlled manipulation of nanoscale defects. Using advanced fabrication techniques such as electron-beam lithography and epitaxial growth, it is possible to create customized material architectures with atomic precision. This capability not only makes it possible to control the density and distribution of Schottky defects, but also to exploit their effects to design high-performance devices and systems, such as ultra-sensitive sensors, low-power electronic devices and highly selective catalysts.

In addition, the combination of advanced experimental techniques and numerical modeling opens up new perspectives for exploring the properties of Schottky defects under extreme conditions, such as high-temperature and high-pressure environments. Using atomistic simulations and sophisticated laboratory experiments, researchers can study the behavior of defects under realistic conditions, providing a better understanding of their stability, mobility and influence on material properties. This information is crucial for designing materials capable of withstanding extreme environments and operating reliably under harsh conditions.

Future research in these fields could therefore lead to significant advances in our ability to design materials with tailor-made properties, by fully exploiting the unique characteristics of Schottky defects at the nanoscale. By integrating knowledge from nanotechnology and advanced experimental techniques, researchers will be able to develop innovative new materials and technologies that meet the growing needs of modern society.

6.3 Potential new applications

An in-depth understanding of Schottky defects opens the way to many potential new applications in a variety of fields, offering innovative possibilities for materials engineering and the development of cutting-edge technologies.

In materials engineering, the ability to control and manipulate Schottky defects opens the door to the design of tailor-made materials for specific applications. For example, by adjusting the density and distribution of defects, it is possible to create materials with optimized electronic, optical or catalytic properties. These tailor-made materials could find applications in fields as diverse as advanced electronics, selective catalysis and energy conversion.

In addition, Schottky defects can play a key role in the development of emerging technologies such as quantum computing and photonic devices. In quantum computing, for example, Schottky defects could be used as qubits, the units of quantum computation, by exploiting their unique quantum electronic properties. Similarly, in photonic devices, Schottky defects could be exploited to control and manipulate the propagation of light at the nanoscale, paving the way for new communication and information processing technologies.

Future research should therefore focus on exploring these potential new applications, with an emphasis on developing new materials and fabrication techniques to fully exploit the potential of Schottky defects in these areas. By combining experimental, theoretical and computational approaches, researchers can explore the limits of these potential applications and identify the best strategies for realizing them.

In conclusion, although challenges remain in understanding Schottky defects, emerging research perspectives and potential new applications offer fertile ground for future innovation in this field. By continuing to explore the properties and behaviors of Schottky defects, we are able to develop new materials and technologies that will shape the future of materials science and engineering.

Conclusion

Materials science is a constantly evolving field, where the understanding of Schottky defects plays a crucial role. These crystalline imperfections, although often considered undesirable, actually offer immense potential for designing materials with tailor-made properties and developing innovative technologies. Throughout this paper, we have explored in detail the origins, characteristics and effects of Schottky defects, as well as methods for controlling and exploiting them.

From advanced experimental techniques to sophisticated numerical modeling, Schottky defect research spans a diverse range of scientific disciplines and technological applications. From nanotechnology to aerospace, electronics to catalysis, Schottky defects are at the heart of many advances and innovations. Their controlled manipulation opens the way to customized materials with specific properties to meet the growing needs of modern society.

As we continue to explore the mysteries of Schottky defects, we are witnessing an exciting convergence between fundamental research and practical applications. The challenges persist, but the opportunities are also endless. By uniting our efforts across disciplinary boundaries and international collaborations, we are well placed to shape the future of materials science and engineering, fully harnessing the potential of Schottky defects to create a better, more sustainable world.

Glossary :

-Schottky defect: A punctual imperfection in a crystal lattice where an atom is missing its place, creating a gap associated with a neighboring ionic atom.

-Doping: The deliberate incorporation of impurities into a material to modify its electrical, optical or chemical properties.

-Defect engineering: The controlled manipulation of defects in materials to optimize their properties or behavior.

-Galvanic corrosion: A form of electrochemical corrosion that occurs when two dissimilar metals are in electrical contact in an electrolytic medium.

-Materials synthesis: The process of creating materials from raw materials or components, often by controlled chemical or physical methods.

-Heat treatment: A process of controlled heating and cooling of materials to modify their properties and structure.

-Controlled crystal growth: A technique for producing crystals with a regular, controlled crystal structure.

-Thermodynamics: Branch of physics that studies the relationship between heat and other forms of energy.

-Entropy: A measure of the disorder or uncertainty of a system, often associated with the distribution of energy in a thermodynamic system.

-Electron microscopy: Imaging technique that uses electron beams to visualize objects on a microscopic scale.

-Spectroscopy: Study of the interactions between matter and light, often used to analyze the structure and properties of materials.

-Numerical modeling: Use of computer tools to simulate the behavior of physical or chemical systems.

-Multiscale correlation: An approach that integrates data from different spatial or temporal scales to understand complex phenomena.

Reference :

1.   Callister Jr., W. D., & Rethwisch, D. G. (2014). Materials science and engineering: An introduction. John Wiley & Sons.

2.   Dieter, G. E., & Bacon, D. J. (1988). Mechanical metallurgy (Vol. 3). McGraw-Hill.

3.   Kittel, C. (2004). Introduction to solid state physics. John Wiley & Sons.

4.   Hull, D., & Bacon, D. J. (2011). Introduction to dislocations. Elsevier.

5.   Cullity, B. D., & Stock, S. R. (2001). Elements of X-ray diffraction. Prentice Hall.

6   Wagner, C. (1961). Theory of the formation of Schottky barriers. Journal of Applied Physics, 32(2), 215-220.

7.   Bhadeshia, H. K. D. H. (2011). Phase Transformations: Examples from Titanium and Zirconium Alloys. Woodhead Publishing.

8.   Smith, W. F., & Hashemi, J. (2006). Foundations of Materials Science and Engineering. McGraw-Hill.

9.   Kingery, W. D., Bowen, H. K., & Uhlmann, D. R. (1976). Introduction to Ceramics (2nd ed.). John Wiley & Sons.

10.   Cahn, R. W., Haasen, P., & Kramer, E. J. (2001). Materials Science and Technology: A Comprehensive Treatment (Vol. 6). VCH.

11.   Dieter, G. E., & Bacon, D. J. (1986). Mechanical Metallurgy. McGraw-Hill.

12   Van Vlack, L. H. (1964). Elements of Materials Science and Engineering. Addison-Wesley.

I want morebooks!

Buy your books fast and straightforward online - at one of world's fastest growing online book stores! Environmentally sound due to Print-on-Demand technologies.

Buy your books online at
www.morebooks.shop

Kaufen Sie Ihre Bücher schnell und unkompliziert online – auf einer der am schnellsten wachsenden Buchhandelsplattformen weltweit! Dank Print-On-Demand umwelt- und ressourcenschonend produzi ert.

Bücher schneller online kaufen
www.morebooks.shop

info@omniscriptum.com
www.omniscriptum.com